Erverton Cezário de Freitas
José Deomar de Souza Barros

Urban solid waste

Erverton Cezário de Freitas
José Deomar de Souza Barros

Urban solid waste

management, management, social and environmental impacts and mitigating measures

ScienciaScripts

Imprint
Any brand names and product names mentioned in this book are subject to trademark, brand or patent protection and are trademarks or registered trademarks of their respective holders. The use of brand names, product names, common names, trade names, product descriptions etc. even without a particular marking in this work is in no way to be construed to mean that such names may be regarded as unrestricted in respect of trademark and brand protection legislation and could thus be used by anyone.

Cover image: www.ingimage.com

This book is a translation from the original published under ISBN 978-620-3-46756-7.

Publisher:
Sciencia Scripts
is a trademark of
Dodo Books Indian Ocean Ltd., member of the OmniScriptum S.R.L Publishing group
str. A.Russo 15, of. 61, Chisinau-2068, Republic of Moldova Europe
Printed at: see last page
ISBN: 978-620-3-63823-3

SUMMARY

With the expansion of cities and population growth enabled the exaggerated consumption of industrialized products, thus increasing their accelerated disposal. Given the advancement of technologies over the years, the products have been changing their composition so that their disposal in inappropriate places causing contamination of the environment. Within this context, the reality of some Brazilian municipalities, especially in the countryside cities, still discard their waste improperly causing environmental damage and loss of quality of life of the population, thus constituting one of the main problems to be solved today. In this sense, several studies have been presented in relation to this problem seeking improvements and developing strategies that can mitigate or at best solve the problem of solid urban waste management. In view of this reality, the present research aimed at analyzing solid waste management by means of sustainability indicators in the City of Uiraúna-PB. The study in question contains both quantitative and qualitative approaches to the disposal of these wastes. The methodology employed for the development of the research was based on Castro (2016) following the model adapted by Bento (2020) so that the indicators were converted from qualitative to quantitative to which there is a better understanding of the data collected. The information was collected by applying a questionnaire to groups of different social actors residing in the municipality of Uiraúna-PB. After evaluating the data, it was found that among the thirty-seven indicators evaluated, twenty were considered favorable and thirteen unfavorable, but it is important to emphasize that most of the indicators that were highlighted as favorable refer to questions of knowledge and opinion about the management, collection, proper final disposal of waste and how these MSW could affect the population when not conditioned in a regular way. While the unfavorable ones were focused on more relevant issues such as the inexistence of an appropriate place for the final disposal of waste, open air dumps near the residences, and the contamination of the population through this waste. Within this perspective the study highlighted the urgency that the municipality of Uiraúna-PB presents so that there are changes in its management of urban solid waste.

Key words: Population growth. Solid Waste. Solid waste management. Environmental damage.

SUMMARY

1. INTRODUCTION

According to (PNRS) - National Policy on Solid Waste, is characterized as solid waste, all material, substance, object that can be discarded, being originated from human action in conjunction with society, whose final destination proceeds to the solid or semi-solid, as well as gases stored in containers and liquids, of which the specifics ultimately make it impractical to dispose of in public networks such as sewers or bodies of water, or that lack technical and economic solutions in view of the best technology so available (BRASIL, 2010).

For Mancini, Ferraz and Bizzo (2012), since the most remote periods, the materials were already present in the history of mankind, and its relationship with the handling of them helped in their daily tasks, where it can be evidenced in some periods of antiquity as: stone age, chipped stone age, bronze age, iron age. Thus, with the evolution of technologies, man began to have at his disposal the most different materials, which enabled scientific advancement and allowed the development of new technologies.

According to Barbosa (2012), the problem related to solid waste had worsened as of the industrial revolution, where the same ceased to be produced in small quantities to be produced in large quantities, allowing the addition or accumulation of waste in urban areas. The same also emphasizes the rampant production of solid waste thus becoming one of the great challenges to be faced by today's society.

With the growing population increase, the rising economy, the slight urbanization and the rise in community living standards, hastened the production rate of this waste (BERTICELLI; PANDOLF; KORF, 2017). Gouveia (2012) points out that solid waste due to its increase during recent years, had its composition changed, where it began to present synthetic elements and highly harmful to human health and the environment, being found that this was also given to the advancement of technologies added to everyday life.

According to Lima (2017), the uncontrolled increase of the global population is also related to environmental impacts, because the general population tends to maintain a consumerist way of life, triggering a direct result in the disorderly exploitation of natural resources, allowing significant increase in pollution levels, also enabling immeasurable environmental losses being characterized by high waste production.

Thus within this reality, the waste when not treated correctly, can cause environmental damage such as: soil degradation through pollution, contamination of surface water as well as the subsoil by infiltration of leachate and there may be the proliferation of disease agents, visual pollution and bad smell (MUCELIN; BELLINE, 2008).

Thus, Berticelli; Pandolf and Korf (2017) point out the great challenges that municipalities face with respect to this problem that is solid waste, where the public institution has an obligation to provide strategies or measures that are effective to mitigate the effects caused by its accumulation. Unfortunately, the vast majority of municipalities continue to use inappropriate techniques and practices for the disposal of this waste.

According to Cempre (2010), public institutions (municipalities) are responsible for developing the form of solid waste management, which allows having as a means, the mode of collection, providing a viable and appropriate end for these solid wastes, seeking techniques in order to segregate and care for the waste produced in the municipality. Thus, the promotion of awareness campaigns are viable techniques aiming at the reduction of this waste where they are subject to change depending on the location and its peculiarities.

In this aspect, sustainability indicators can prove to be of great importance considering that it can facilitate access to technical and scientific information, in which its main function is to organize data so that they are easy to understand (PEREIRA; CURI; CURI, 2018). Thus, sustainability indicators are of great use and can facilitate the analysis of data related to solid waste management, enabling the creation of strategies and methods for each situation and its specifications (SINGH et al., 2007).

Thus, this research seeks to contribute and seek improvements in waste management in view of the current situation, using sustainability indicators to verify the quality and impacts caused by this waste in the city of Uiraúna-PB; and aimed at analyzing solid waste management through sustainability indicators in the city of Uiraúna-PB.

2. THEORETICAL BACKGROUND

2.1 Environment and sustainability

The incessant search for stability between economic growth and conservation of environmental resources has reinforced the idea of sustainable development (KEMERICH; RITTER; BORBA, 2014).

According to Kemerich, Ritter and Borba (2014) sustainability is a process of changes that occurs gradually, seeking improvement in different aspects and dimensions having the participation of the population as a transforming agent. According to Medeiros, Giordano and Reis (2012), this term "sustainability" was presented in a work called "our common future" which was produced by the World Commission on Environment and Development. The same was exposed at the conference on the environment and sustainable development Rio 92.

For Medeiros, Giordano and Reis, (2012) the need for changes with regard to sustainability led to consider the relationship man / environment, because the inappropriate use of natural resources made by man throughout its evolutionary history has become in general, extravagant and abusive, because it tends to sustain a consumerist way of life. As a result, the available natural resources are not able to ensure the sustainability of ecosystems, where their demand is increasing.

For Zasso et al. (2014) this occurred due to the accelerated evolution of urbanization in view of the new standard of living adopted by modern society (capitalist), which increasingly enabled the estrangement of man with the environment. The same also emphasizes the changes made by man, who modifies according to his needs, disregarding the dynamics of nature directly affecting environmental quality.

Thus Pordeus and Barros (2017) emphasize that the use of environmental resources inappropriately ends up degrading the environment, causing irreversible impacts, which can cause environmental imbalance affecting all ecosystems.

Within this context, according to Sartori, Lantrônico and Campos (2014) sustainability acts as a guidance mechanism instructing how one should act in relation to nature, emphasizing the responsibilities with the environment, allowing the conservation of the natural environment for future generations. Thus Querino and Pereira (2016) emphasize the construction of a mentality towards environmental issues, which integrate new knowledge, transforming traditional scientific paradigms, which collaborates with different specialties and interdisciplinary organization of knowledge for sustainable development.

Thus, for Barbosa (2008), it is important to look for sustainable means that include quality of life within this reality, for a good urban interaction, where it serves as a reference in the urban planning processes.

2.2 Solid Waste

For Mancini, Ferraz and Bizzo (2012), since the most remote periods, the materials were already present in the history of mankind, and its relationship with the handling of them helped in their daily tasks, where it can be evidenced in some periods of antiquity as: stone age, chipped stone age, bronze age, iron age. Thus, with the evolution of technologies, man began to have at his disposal the most different materials, which enabled scientific advancement and allowed the development of new technologies.

Mancini, Ferraz, and Bizzo (2012), also emphasize the complexity that this development took, because the production of these materials occurred in order to meet the needs of the population in view of the fact that the population increased over the years and currently the population density is the highest ever.

Following this sense, these materials that after its use receives the name garbage, can be conceptualized as everything that can be discarded, thrown away or that has no more use (TENORIO; ESPINOSA, 2004). With the industrial revolution the garbage came to be called solid waste, i.e., material that has undergone a long process in its manufacture and as a result underwent changes in its chemical and physical composition, so it ceased to be produced on a small scale to be produced in large quantities, enabling its accumulation in urban areas (BARBOSA, 2012).

These solid wastes according to Tenorio and Espinosa, (2014) are classified in view of their origin, chemical composition and characteristics and can be called industrial solid waste, urban, health services, ports, airports, road and rail terminals, agricultural, radioactive and rubble. According to Castro (2014) this accumulated waste can, if not disposed of properly, affect the quality of life of a population. Thus, it stands out for degrading the environment and being a means of disease proliferation and contamination of natural resources such as soil and water reserves.

For Santos and Dias (2012) this problem related to municipal solid waste is treated with urgency given the circumstances of the municipalities and metropolitan regions, where the precariousness and neglect of the public power contributes to the mismanagement of MSW.

According to Tenorio and Espinosa (2004), some data portrays the distribution of this waste among the final destination sites, highlighting the great deficiency Brazil presents in as much as basic sanitation is concerned. The data provided by the Brazilian Institute of Geography and Statistics - IBGE concerning basic sanitation in 2000 point out that of the 230 thousand tons produced per year, 22% are destined to open air dumps or dumpsites, considering that 75% of these residues are deposited in sanitary landfills and controlled places, but the amount of residues deposited in inadequate places is still high.

2.3 Sustainability indicators

For Van-Bellen (2005) sustainability indicators can be conceptualized as a mechanism that simplifies information, quantifying it in an organized and consistent manner within a perspective that reflects the reality according to certain phenomena. The same author also highlights the emergency that the environmental theme exposes, since the lack of understanding on the subject is firmly linked to human activities in relation to natural resources.

According to Silva, Candido and Ramalho (2012) this idea of sustainability makes it possible to reach issues related to the environment and urbanism to which it was solidified in the period of the 90s, but the use of this concept is far from being understood more broadly. In view of this Kemerich, Ritter and Borba (2014), highlight the demand and the growth of these sustainability mechanisms by the government, federal institutions (universities, research centers), non-governmental institutions around the world, where there were several conferences, lectures, research promoted by these institutions.

According to Lima (2017), sustainability indicators can also be characterized by gathering both qualitative and quantitative information, but some authors prefer to mention only quantitative data which they believe are better for analyzing sustainability due to the scarcity when compared to numerical indicators.

According to Lima (2017)   , these indicators were created with the goal of gathering information about a certain reality, with a view to simplifying and summarizing them, leaving only the relevant and significant information.

Thus, according to Pereira, Curi, and Curi (2018), with the help of these sustainability indicators directed especially to solid waste, it becomes of great importance for enabling elements that can guide decision making in various ways, and can simplify information helping in their management, and assisting in the identification related to socio-environmental aspects.

2.4 National solid waste policy

According to (PNRS) - National Policy on Solid Waste, the term solid waste can be defined as: material, substance, object that can be discarded, having its origin in human action, to which its final destination proceeds to the solid or semi-solid, being finally unfeasible disposal in the public network (BRASIL, 2010).

Thus, in the face of several negative points in relation to the impacts caused by the addition or accumulation of solid waste, there was the need to create public policies in order to ensure the management and proper disposal of the same. Within this context was implemented the PNRS instituted by Federal Law No. 12.305/2010, which was considered the starting point for the regulation of solid waste (LIMA, 2017).

Following this line of reasoning the PNRS encompasses a systemic field of vision integrating aspects and dimensions related to solid waste management, among which, we can mention environmental, social, cultural and also public health aspects (CASTRO, 2014). Thus for Gonçalves (2012) The National Solid Waste Policy (PNRS) are laws that interfere in the lives of all individuals that make up the society, from the social living to the most varied sectors, such as the industrial sector. The same also emphasizes that the PNRS, as well as any other law fits the desires of society, where in the future there may be changes if necessary

Thus, according to Law No. 12,305 of August 2, 2010, art. 4, the National Solid Waste Policy encompasses a set of objectives, mechanisms, principles, guidelines, targets and actions, instituted by government agencies (federal government, municipalities or individuals), so that they can be worked on individually or jointly, aiming at full cooperation and progress with regard to the proper management of solid waste. (BRAZIL, 2012).

3. METHODOLOGICAL PROCEDURES

The research was conducted in the municipality of Uiraúna - PB, in the period from April 11 to September 19, 2019.

3.1. Characterization of the study area

The municipality of Uiraúna - PB is located in the Brazilian Northeastern semi-arid region, west of the State of Paraíba, in the Microregion of Cajazeiras, which is part of the Mesoregion of Alto Sertão Paraibano. Having its limits to the north with the Potiguar municipalities of Luís Gomes and Paraná; to the (South), to the east and west with the following municipalities of Paraíba: São João do Rio do Peixe, to the (South); Vieiropólis, to (East), Joca Claudino and Poço de José de Moura to the west (IBGE, 2020).

Its position is demarcated by the geographical coordinates with latitude 06º 31' 03" S, and Longitude: 38º 24' 28" W, has an altitude of 301 meters above sea level. Its territory is 293.182 km² and has a population of 15,300 inhabitants (IBGE, 2020).

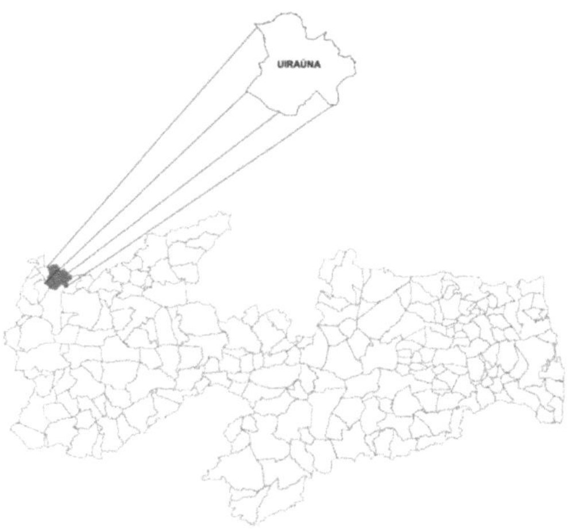

FIGURE 01. Location of the municipality of Uiraúna-PB within the Paraibano Territory. Source: CPRM (2005)

3.2. Research Classification

From the standpoint of its nature, the research was characterized as applied research, in which, according to Zanella (2013), it aimed to generate knowledge for practical application aimed at solving specific problems and local interests. From the point of view of problem approach it is said as quantitative research, according to Zanella (2013), this type of research considers that everything can be quantifiable, which means translating into numbers opinions and information to classify and analyze them, using resources and statistical techniques. As for the objectives, the research is descriptive, according to Prodanov and Freitas (2013) descriptive research aims to describe the characteristics of a given population or phenomenon or the establishment of relationships between variables. It involves the use of standardized techniques for data collection: questionnaire and systematic observation. And as for the technical procedures to be performed, the research is considered a case study, according to Prodanov and Freitas (2013) the case study involves the deep and exhaustive study of one or a few objects in a way that allows its broad and detailed knowledge.

3.3. Indicators used

The indicators used in this study are similar to those used in the methodology of Castro (2016), where the same qualitatively classified the indicators in favorable, unfavorable and very unfavorable. At the same time, in order for the data to be easily understood, it was decided to classify them quantitatively, assigning values of 5, 3, and 1, respectively. In this way, at the end of the survey, the evaluations are added up. The present work also sought to highlight the indicators in favorable and unfavorable thus giving new values, 1 and 0 respectively, following the model used by Bento (2020) where the same was necessary to facilitate and understand the variants.

3.4. Research Subjects

The subjects that were investigated in the research are people who live in the municipality of Uiraúna - PB, located in different urban areas. The information that was collected made it possible to obtain more reliable data, due to the fact that it takes into account a wider range of locations and thus enables more coherent data to the reality studied, providing research with greater performance and contributing to an analysis of the quality of waste management in the urban space and later a reflection that seeks to contribute to the improvement of the final disposal of solid waste in that location.

3.5. Population, sample and sampling.

The subjects that were surveyed are residents of different localities of the municipality in question. Considering that the municipality has a large number of inhabitants, the sampling technique was used, where these residents were randomly selected based on their place of residence, where sixty-six social actors residing in the municipality of Uiraúna - PB were surveyed.

3.6. Data Collection

Data collection was carried out through the application of questionnaires. According to Silva and Menezes (2005) these expose an ordered series of questions to be answered by the research respondents. Their use facilitated the obtaining of a large amount of information in a short period of time during which the research was developed.

3.7. Data Analysis

Soon after the data evaluation, the indicators were classified as favorable, unfavorable, and very unfavorable, assigning different values for each one according to the model used by Castro (2016), and suggestions were developed to improve tailings management in that location.

4. RESULTS AND DISCUSSIONS

The research included a total of sixty-six respondents, all from different neighborhoods of the municipality of Uiraúna PB, of which 63.6%, which corresponds to 42 respondents were male, while 36.4%, which corresponds to 34 respondents were female, a result similar to that of Bento (2020), in which the study conducted in the municipality of Nazarezinho-PB shows a higher percentage with men than with women. However, the same also points out that studies conducted in the city of Cajazeiras-PB, the percentage of respondents had as a larger group, females, highlighting the business hours as a possible factor influencing the result. Thus, according to the results, in the locality, males were more expressive.

Regarding the age range, it was found that most respondents were aged between 30 and 40 years, with a percentage of 36.4%, corresponding to 24 respondents, 20 to 30 years, had a percentage of 34.8% representing 23 respondents, as the respondents with up to 20 years the percentage was 6.1% equivalent to 4 respondents, with 40 years or more, it was found the percentage of 22.7% corresponding to 15 respondents. In research conducted by Barros, Chaves and Farias (2014), the predominant average age of heads of household was 40 to 60 years, contrary to the data obtained. However, data recorded in the municipality of Nazarezinho-PB by Bento (2020) found that the age range of household heads varies, being over 25 years, thus demonstrating the similarity between the data obtained, which shows that the heads of household over 25 years old have greater predominance, the agreement between data denotes a reality not only territorial, but regional in the semi-arid Paraiba.

As for the level of education, it was found that the largest portion of respondents are people who have complete high school education with 34.8% corresponding to 23 of the respondents, while respondents with complete higher education amount to 25.8%, with complete elementary school 15.2%, incomplete high school 10.6%, with incomplete higher education 9.1%, incomplete elementary school 4.5%. In this sense, the results when compared to those of Vasquez, Barros and Silva (2008) show a higher level of education in relation to respondents who have completed high school and also elementary school. This also happens with the data presented by Barros, Chaves and Farias (2014), in which the data obtained show a better level of education. With Bento (2020) it is possible to observe the similarity of the results in question, because it shows that over the years there has been an improvement in the educational level of community members.

4.1. Knowledge about solid waste by the population

It becomes important the way in which the human being interprets the environmental conditions which surround him, thus allowing the construction of knowledge in order to understand the environment in which he is inserted. In this sense, the way man perceives everything that occurs around him is linked to a perceptual process or an information system that connects man directly to people, objects, events of everyday life (QUEIROZ, 2011).

Pereira and Querino (2016) points out that human beings are endowed with intelligence and intellect, in which man is aware of these materials and through the same subsidize and provide for their survival extracting the maximum of the natural environment, imposing its will to provide welfare and comfort.

Thus having perception and awareness of this waste, the human being interacts, responds and perceives the different changes in the environment, instigating a broader understanding in relation to the dynamics between man-environment (PEREIRA; QUERINO, 2016). Most of the public, when asked about their knowledge of what solid waste is, the result was positive, corresponding to 56.1%. However, 43.9% were unaware of what this waste is.

Oliveira, Santos and Viana (2016) explains that there are still many doubts on the part of community members about what is solid waste, while highlighting the importance of public policies and the development of actions aimed at environmental education in the locality. Thus, considering that most respondents are aware regarding the knowledge of this waste, it is said to be **FAVOURABLE**, because the percentage reaches a significant figure in relation to what was asked.

4.2. Solid waste produced

The waste production can vary according to the population, taking into account that this variation can be caused by factors such as: periods of the year, social class, daily habits, consumption of semi-prepared or industrialized products, localities or region where they live (OLIVIERA; SANTOS; VIANA, 2016). According to the respondents, when asked about the production of solid waste, 72.7% replied that the largest production of waste produced corresponds to inorganic waste, 27.3% claimed to be organic waste. According to Rodrigues (2010), this is explained due to the high availability of whole or semi-prepared food products, so there was a reduction in the disposal of organic waste, in contrast to the increased incidence in the disposal of inorganic waste.

Whereas inorganic waste consists of: paper, glass, cardboard, plastic among others. Ciquetti (2004) points out that the main environmental problems are related to this waste, highlighting non-degradable product packaging, disposable or toxic products. Thus, it is considered **DISADVANTABLE**, because in relation to the large production of inorganic waste it becomes worrisome if there is not an appropriate final destination for this urban solid waste in the municipality.

4.3. Urban cleaning execution

The great challenge for expanding cities is precisely to provide quality urban cleaning, which does not consist of just removing waste from the main roads, streets and buildings, but rather enabling a viable and appropriate disposal for this waste (GIACOM-RIBEIRO; MENDES, 2018).

Cempre (2010) points out that the municipalities are responsible for managing the waste produced, thus being responsible for providing strategies and methods that can enable a better disposal of these wastes.

Based on the results obtained, 75.8% of respondents stated that the execution of urban cleaning is the responsibility of the municipal government, while 4.5% stated that urban cleaning was performed by others, and 19.7% of respondents answered that the city hall in agreements with other agencies. According to the data analyzed, Giaccom-Ribeiro and Mendes (2018) reports that the municipalities do not have the technical and financial apparatus to deal with the problems related to waste management. Thus, it becomes **DISFavorable,** because even if the institution is responsible for making the final disposal of this waste feasible, it does not have enough resources to provide the management of this waste effectively, making the problem even more worrisome.

4.4. Garbage collection within the municipality

Garbage is defined as everything that is disposable, has no more use, useless or undesirable, arising from human action that varies from solid or semi-solid (RICHTER, 2014). For Lima (2018), this waste on public roads can represent a high risk to the population, given that its accumulation favors the appearance of disease-transmitting agents. Thus, the collection of solid waste becomes an essential factor in preventing possible sources of disease in the municipality.

With the data analyzed, 66.7% of the participants declared that the garbage collection covers the entire municipality, while 33.3% answered no. Considering the result obtained, it is said to be **FAVOURABLE,** since the collection of this waste occurs throughout the municipality.

4.5. Charging for cleaning services

Leite (2006) points out that urban cleaning is one of the largest investments made by Brazilian municipalities, whereby the collection, transportation and final disposal of such solid residues are within the services supplied by municipalities. In which the collection of cleaning services by municipalities is unconstitutional, having been rejected by the supreme federal court.

According to the data collected, 78.8% of the participants emphasize that the services provided by the municipality do not require payment by the population, since the municipality bears the costs related to the collection and final disposal of this waste. 15.2% emphasize that they do, 3.0% affirm that a fee is charged, and another 3.0% answered that it is charged along with the IPTU. Within this context, according to the data analyzed, it is said to be **FAVORABLE** because the cleaning service is accessible without charging a fee to all inhabitants of the municipality.

4.6. The population's knowledge about selective collection

Selective collection becomes an important factor with regard to environmental preservation and sustainability. The same consists of the separation of these wastes, separating them organic from inorganic, thus allowing a viable and appropriate disposal. The materials that are collected and recycled are mostly paper, glass and metals (RICHTER, 2014).

According to Marco (2014), selective collection is a constitutional obligation of the government, which enables several improvements for the municipality, among which is the reduction of waste collected and treated in landfills, ease of fundraising, reduced pressure and demand from environmental agencies, among others.

When asked about what is selective collection, 74.2% of respondents pointed out that they had knowledge about selective collection, while 25.8% replied that they did not. This data is similar to the studies of Bento (2020) in which, in the municipality of Nazarezinho-PB the percentage of the population regarding the knowledge of what would be selective collection reached a percentage of 80%.Given this reality, according to Richter (2014), to be successful with regard to the selective collection, it is necessary to raise awareness of the population in relation to the waste generated, and they can separate them before the collection.

In this way the population's awareness becomes the first step to be taken, bearing in mind that the population has a fundamental role in the process. In this way it becomes **FAVORABLE**, because having pointed out that the population has a fundamental role in the awareness process, it contributes to a more sustainable environment by increasing its quality of life.

4.7. Period in which the collection is performed

According to the complementary law Nº 234/90 article 2, is characterized as urban cleaning service: collection, transport and final disposal of waste from the population is they ordinary, household or special (BRASIL, 1990).

In agreement with Lima (2018) the collection in the municipality is done directly in the neighborhoods of residents by cleaning agents in which these wastes are transported by trucks having their final disposal in the dump.

According to the data obtained, when asked about the period in which the collection service passes by to do the collection, 74.2% of the interviewees affirmed that the collection is done weekly in the city, daily and alternating both with 9.1% and eventually 7.6% respectively.

Thus it is classified as **FAVORABLE**, because the collection of solid urban waste in the municipality is done regularly, being done weekly and covering all neighborhoods in the municipality in question.

4.8. Degree of satisfaction with waste collection

The collection process occurs in stages, from the time the collection vehicle leaves the parking lot of its garage, covering the path that the vehicle spends collecting the residues from the places where they were conditioned, to the disposal places, until returning to the starting point (CUNHA; FILHO, 2002).

According to the data obtained, 36.4% of those interviewed considered the quality of collection good, the second most obtained answer was that the quality was regular, with 34.8% of the interviewed population. 19.7% considered the quality of collection to be excellent, while 9.1% considered it bad.

In this way, it is considered **FAVORABLE**, because analyzing the results, the collection covers the entire municipality, even if in some cases there is no adequate disposal or effective collection.

4.9. Knowledge of the population about the final disposal of waste in the municipality

According to Vieira and Garcia (2012), in Brazil it was found that in 2006 around 51.1 million tons of tailings were generated, bearing in mind that 18% of the total were metals, paper, plastics and glass that were recycled.

Regions north and northeast, had the highest rates of waste with final disposal mostly to open air dumps, about 89.3% and 85.5% had this final destination (SHIMTZ, 2012).

The problem in question becomes a major challenge for Brazilian municipalities that still mostly throw their waste in open air dumps, where data obtained by the Brazilian Institute of Geography and Statistics show that 50.8% of waste produced in the country has its disposal destined for dumps in which it can be seen that its greatest aggravation is in the northern region of the country in which municipalities throw about 59% of their waste in open air dumps (COSTA et al., 2016)

In the municipality of Uiraúna, when asked about where the waste produced goes, 72.7% responded that this waste is disposed of in dumps, 13.6% of respondents said that this waste was sent to landfills, 3.0% said it was incinerated, 10.6% do not know the final destination of this waste. In agreement with Costa et al. (2016), the improper disposal of this waste can compromise the balance of the ecosystem, causing pollution of water, air and soil, enabling the proliferation of agents that cause diseases.

Thus, with the data collected, it is said, **DISFavorable**, because in the municipality in question, there is no sanitary landfill and its waste is sent to landfills.

4.10. How the residues produced at the residences are stored

Considering that in the waste management cycle, after the production and use of waste, storage occurs, it can be classified into three stages: adequate, inadequate and no packaging. Thus, adequate would be the use of plastic bags, buckets or cans with lids, while the inadequate way to store this waste are: buckets without lids, wooden crates and cardboard boxes (BENTO, 2020).

Thus, according to the interviewees when asked about the way they store the waste produced in their homes, 62.2% responded that they store waste in a plastic garbage can with a lid, while 36.4% of respondents said they pack the waste in a plastic bag. In this aspect is said, **FAVORABLE**, because it is considered that residents store the waste properly.

4.11. Separation of waste types

The separation of waste is in fact a little used activity by the population that packs its waste without any kind of segregation, that is, the waste is stored in a way that papers, plastics, metals, and organic waste share the same container in which it will be collected by the public cleaning professionals.

Within this reality, the collectors who work in the door-to-door collection are subject to various types of risk. When there is no proper packaging of garbage, the collectors as they perform the collection may suffer cuts by metal, glass, needles, among other materials precisely because citizens do not properly separate (KLEIN; GONÇALVES-DIAS; JAYA, 2018)

Based on the data observed, when asked whether they separate their household waste, 54.5% of respondents said they do not separate, 45.5% said yes, they segregate the waste produced. Thus this aspect is considered **DEFAULT**, given that most do not perform the separation of waste and can harm both cleaning professionals as well as the environment in which they operate.

4.12. Reuse of waste by respondents

The reuse of waste became of great relevance in the post-war period, in which inhabitants **of** countries that were affected by the war began to reuse materials by transforming them in order to subsidize their survival (SOUZA; MELLO, 2015). In Brazil, the idea of reusing or recycling waste became well known due to ECO-92. From there the world became aware of issues related to conservation and preservation of the environment (SOUZA; MELLO, 2015).

With this perspective, recycling helps to reduce the amount of waste collected, facilitating the work of the collection professionals, as it contributes to maximizing the family income of the collectors who work in the recycling of this waste.

In this aspect 74.2% answered that they do not reuse the waste consumed in any way, while 25.8% said they reuse, but mostly plastic and glass. Within this reality it can be seen that most of the population does not reuse waste, highlighting this aspect as **DISFavorable**.

4.13. Waste separation for recycling

For Silva and Noleto (2004), it is important to disseminate knowledge about the proper form of waste separation, so that it is extended not only to the homes of residents, but also to restaurants, bars, leisure areas and so on, making it disclosed daily so that the population is always on the alert. In this case, the absence of means that can enable communication between the population and the public sector about appropriate forms of management of this waste can consequently lead to the population packing it in the wrong way.

According to Matos (2013), the inadequate disposal of this waste in clandestine deposits can cause environmental damage and affect the quality of life of the residents in that location.

In this question the negative statements had a majority of 57.6%, while those who separate waste for recycling correspond to 42.4%, so it was possible to show that the largest portion of the population does not separate the waste produced, while the percentage that makes the segregation of these wastes performs the separation between organic and inorganic waste. After analyzing the data collected, this question is considered **DEFAULT**.

4.14. The use of more sustainable products

In a social context in which the population's unbridled consumption allows for the uncontrolled generation of waste, it becomes necessary to create strategies to mitigate the effects provided by the high production and inadequate disposal of this waste.

This is further aggravated due to the scarcity of knowledge by the population regarding environmental education tied to a society disorganized in social and economic issues (PORDEUS et al., 2019)

In this aspect it is necessary a vision directed to the moderate consumption of environmental resources worrying about the current good and future consumption, this way there must be social/economic welfare, democratic citizen participation and environmental sustainability (VICENTE; BERTOLI; RIBEIRO, 2016).

Given this reality, the use of products that are less harmful to the environment can promote equity in consumption/waste generation, also adopting forms of reuse, because according to Vicente, Bertoli and Ribeiro (2016), adopting techniques such as recycling, the first step is taken to combat unbridled consumerism destabilizing the consumption chain and inadequate waste disposal.

For this question, it was analyzed that 57.6% of respondents use products without worrying that they may harm the environment, 42.4% try to use less aggressive products to the environment, thus it was possible to verify that the population in its majority use products without worrying whether or not they can harm the environment, thus characterizing this question as **DISFavorable**.

4.15. Degree of satisfaction with selective collection

The selective collection is a systematic service, which works in a way to segregate the collected material, while facilitating and reducing the waste taken to disposal sites, making the material valued, promoting its reuse within the production environment (LEITE, 2006).

Thus the associations and cooperatives of collectors become an important means to assist in the communication between the sectors of public power, private and population in general, emphasizing an integrated approach regarding the selective collection, so that the sectors provide conditions for these professionals to act effectively (KLEIN; GONÇALVES-DIAS; JAYA, 2018).

Klein, Gonçalves-Dias, and Jaya (2018), also highlight the situation of these associations and cooperatives, in which their activities in some Brazilian municipalities act in an informal manner, even being governed by the PNRS a greater portion still perform their functions irregularly or individually.

In the Municipality of Uiraúna - PB the selective collection is performed weekly where it follows the schedule of waste collection in their respective days, Mondays, Wednesdays and Fridays, being held by the association of collectors of the municipality. For 47% of the interviewees, the selective collection is performed effectively, classified as good, for 27.3% classified as regular, 22.7% replied that the selective collection is great, for the other 3.0% classified as bad. Thus it is said **FAVORABLE,** where the selective collection has been performed effectively meeting all the demand of the municipality in question.

4.16. Final disposal of hospital waste in the municipality

Hospital waste or health service waste is defined as waste from activities performed in health units such as dental clinics, hospitals, veterinary clinics, outpatient clinics, health clinics, etc. (CAFURE; PATRIARCHA-GRACIOLLI, 2015).

These wastes, due to the fact that they cause contamination, can become a public health problem, and can also compromise the quality of life of the population that are in direct or indirect contact with these hospital wastes (CAFURE; PATRIARCHA-GRACIOLLI, 2015).

According to the health department of the municipality, hospital waste is disposed of by the company that has an agreement, and ensures that this waste is properly disposed of. So for the interviewees when asked about the destination of this waste, 77.3% replied that they are unaware of the disposal of this waste, and 22.7% said they know where this waste goes.

Although most of the interviewees do not know where the hospital waste goes, the municipality packages and makes possible a viable and appropriate destination for this waste, thus the same is said to be **FAVORABLE,**

4.17. Knowledge about the dangers of hospital waste

For Cabral et al (2015), solid waste, in general, without the necessary care causes great damage to the environment and consequently the loss of quality of life of the population, but when it comes to hospital waste the risk in relation becomes potentialized, because it has a high biological risk, ie they are capable of spreading diseases.

With all this it is worth noting that standards, guidelines and laws are established for the proper management of these wastes, however there is still inadequate packaging of these wastes (CAFURE; PATRIARCHA-GRACIOLLI, 2015).

Thus according to Camargo et al. (2015), it is of great importance the responsible control of this waste enabling techniques and solutions so that it can have a viable and safe collection, storage place, treatment of proper final disposal.

As for batteries, Bento (2020) explains that this waste, when not properly stored, can harm the environment considerably, and may result in the release of its toxic components into the natural environment, thus contaminating natural resources and decreasing the quality of life of humans.

For this indicator 86.4% of respondents demonstrated knowledge about the dangers of these wastes to health, both human and environmental, those who do not know represent a percentage of 13.6% of respondents. This indicator is considered **FAVORABLE**, because the population has knowledge about the risks to human and environmental health provided by this waste.

4.18. Popular acceptance of environmental awareness campaigns

The environment has always been a topic that has generated several discussions in various sectors, thus reflecting on aspects such as economics, politics and also on the way of life of contemporary society. Within this reality, it is true that there is an increase in the number of associations, institutes, governmental and non-governmental organizations that seek to implement awareness policies, while also aiming to educate about environmentally appropriate activities (CAVALCANTI, 2013).

Thus through the dissemination of knowledge these bodies corroborate for the dissemination of campaigns and lectures, using technologies in order to encourage the population about the importance of the theme (CARVALHO; ESTENDER, 2017)

When asked about their opinion regarding the development of environmental awareness campaigns, the interviewees were mostly in favor, finding 89.4% of cases, against 10.6%, considering then **FAVORABLE** for this indicator, since the population was receptive to possible guidelines that can be adopted through awareness campaigns.

4.19. How the population would like to receive this kind of orientation

Following the model of Bento (2020), in this indicator and for the next to follow, its assessment was different from the others performed, in this aspect it was accepted more than one alternative chosen by the respondent, in which there were the guidelines in specific which the same would like to receive, thus the respondents could have a range of choices in the questionnaire and thus could choose the most feasible alternatives to receive this type of guidance. To verify the percentages of the aforementioned indicator and the next ones to follow, the sum of the respondents' statements was performed, where the average of alternatives chosen by the public was between 2 and 3 alternatives, thus through the basic calculation of simple rule of three it was possible to make it simpler to understand.

In this question 62.12% agreed that the most feasible way to receive this type of guidance would be through visits to their homes, while community meetings and information by radio both had a similar percentage with 16.67% of the statements, already 15.15% prefer that these guidelines are linked through pamphlets or posters, 4.55% of respondents prefer dissemination by newspapers. It is considered **FAVORABLE** in question, since the population is in agreement and accepts the forms of disclosure and guidance on environmental issues.

Within this reality, the technologies used for the dissemination and propagation of material related to environmental issues enable the development of society, making them seek more sustainable means of avoiding waste, promoting the dissemination of actions such as recycling, selective collection, so that the population can contribute effectively in the conservation of the environment (CARVALHO; ESTENDER, 2017).

4.20. Types of materials consumed in the households

Plastics are materials derived from petroleum, which is a raw material considered cheap, with durable life and quite versatile facilitating the development of products for society, however these materials can cause serious damage to the environment and human health due to chemical additives used in its manufacture (OLIVIEIRA, 2012).

When asked about the type of material produced in their homes, organic material was the one that presented the greatest amount in the homes of the participants, adding up to 53.03%, second comes the plastic with 50%, other 48.48% said it was the paper. Thus according to the data, this indicator becomes **DISFavorable**, because although organic waste is produced mostly in the homes of respondents, it is found that inorganic waste is consumed without recycling or reuse of the same.

4.21. Types of garbage produced within the municipality

According to Rodrigues and Menti (2016), a consumerist way of life, added to the modernization of technologies, allowed products to become obsolete, increasing unbridled consumption, consequently increasing the disposal of this waste by the population itself.

When asked about the waste produced within the municipality, 65.15% said that the most produced type of waste within the municipality would be public waste, waste from sweeping activities of squares and other forms of public cleaning, second with 31.81% household waste, 15.15% responded commercial waste, agricultural waste and rubble had similar percentages, both with 9.09%, another 4.54% declared industrial waste. This indicator becomes **FAVORABLE**, because the garbage coming from the population in larger quantities within the municipality are waste that can be recycled thus reducing the waste collected if there is selective collection within the municipality.

4.22. Main problems caused by inadequate waste disposal according to the population

Since ancient times man   has been suffering with the negative effects of   these wastes, due to inadequate management, which can be evidenced in the eighteenth and nineteenth century when there was rapid growth in the industrial sector and also urbanization (SIQUEIRA; MORAES, 2009).

Soares (2017) points out that the lack of resources to invest in waste management would be the main problem for its inadequate management, this also tied to the lack of effort and interest on the part of the public sector to take more effective measures, in addition to the lack of collection from the population. In this context, the garbage when accumulated allows the generation of bad odors besides being the precursor of several types of diseases. When not properly treated it becomes a potential risk to the quality of life of society and the environment (SILVA, 2015).

In view of this indicator, 74.24% of the statements were directed to concern about the health of the population, however, it was not evidenced that any popular has been contaminated by diseases arising from the presence of the dump, 24.24% responded that the presence of this waste compromised mainly the quality of the environment and its natural resources, 12.12% questioned the generation of bad odor that these wastes caused. Thus, the indicator is taken as **FAVORABLE**, because people are concerned about their health and have knowledge of the risks that these places represent to their health.

4.23. Main parties responsible for the damage caused to the environment

With the exponential growth of the population associated with industrial development in the eighteenth century was intensifying the environmental impacts. As a result, there was an accumulation of garbage in the main avenues and streets, making possible the emergence of diseases and consequently leading to the deaths of thousands of people (GIACCOM-RIBEIRO; MENDES, 2018).

In current times, the environment is increasingly being affected by human activity. According to Soares (2017), the cause of these impacts is precisely due to the way of life that man exerts, aiming at high economic power and consumer goods.

According to the interviewees, the greatest responsible, with 66.7% of the indications is the population itself, and many of the interviewees criticize the habit of people throwing their waste in inappropriate places, especially in vacant lots, making these places foci of possible diseases. With 18.2% they blame the city hall, because being a public body, the sector should be more committed to providing a quality service to the population. As the most responsible for the degradation of the environment, both the agricultural and the commercial sector had 6.1% of the indications, and 1.5% pointed to the industrial sector as being the most responsible. According to the data analyzed, this indicator is said to be **FAVOURABLE**, because the population is aware of their inappropriate habit at the same time that recognizes the environmental impacts caused by the improper disposal of such waste, in addition to recognizing that the sectors could provide a better service to the population.

4.24. Considerations about the Quality of the municipality's current waste disposal site

According to Sousa, Ferreira and Guimarães (2019), the open air dumps, popularly known as dumps, correspond to a place where waste is discarded indiscriminately, without planning and no control with respect to environmental impacts. The same also points out that the disposal of waste in these places is not only inadequate but also illegal, where he emphasizes that the appropriate place for the correct disposal would be the sanitary landfills, which are controlled and environmentally appropriate places.

In this sense we can conclude, according to Gomes et al (2015), that landfills are the most suitable places for the disposal of these wastes, where they have quality in relation to techniques and methods used for the control of these wastes, thus reducing environmental impacts.

When asked about the quality of the place where waste is disposed of in the municipality, 54.5% answered that they are aware that the disposal site is not the most appropriate, because the place is not safe and at the same time it may pose a risk to human health and also cause environmental damage. 45.5% are unaware of the danger that these sites represent to the quality of life and also to the environment. Thus, this indicator is presented as **DEFAULT**, because most of the population knows that the site needs to be improved, however it is still with its inadequate use.

4.25. Implementation of a sanitary landfill system according to the interviewees

According to Portela and Ribeiro (2014) landfill is the place where waste from human activities is properly directed, so that the disposal of these wastes is controlled and planned, causing the minimum impact to the environment.

With everything the landfill seeks engineering methods and techniques to make possible its construction in an area with small space and reduce as much as possible these residues, having quality and being very favorable regarding the cost/benefit (CONDE; STARCHIW; FERREIRA, 2014).

Sousa, Ferreira, and Guimarães (2019) point out that despite having more sustainable ways to dispose of this waste, most Brazilian municipalities still do not have controlled sites for waste disposal. Small municipalities are those that still suffer from this type of situation, for not having technical teams capable of making feasible and creating more effective measures.

According to the respondents' opinion, regarding the implementation of a sanitary landfill system, 40.9% of the participants answered that its implementation would be great, since the waste directed to the landfill would be treated properly and responsibly, 34.8% said it would be good, the introduction of such a system. Considering that most of the population see as an improvement the installation of the landfill in the municipality, this indicator is classified as **FAVOURABLE**.

4.26. Opinion about waste disposal in landfills

The landfill is considered the most appropriate place for solid waste disposal, where it consists of sanitary techniques for better management of the waste produced, as well as the treatment of gases and also the leachate from the accumulation of waste (PORTELA; RIBEIRO, 2014).

According to Ritcher (2014) the role that selective collection has within sustainability is of great value, because it helps in the processes of environmental preservation, given that the use of selective collection has a significant impact on the reduction of solid waste, besides making that waste return to the consumption environment.

Thus the sanitary landfill aggregated with the selective collection system enables the useful life of the landfill, while allowing the introduction of environmental education reinforcing the importance of an ecological awareness allowing greater participation of the population (PORTELA; RIBEIRO, 2014).

Considering the opinion of the participants, when asked about the type of waste that should be discarded in landfills, most of them, 34.8% answered that hospital waste needs a different destination from the others, because the handling process requires more specialized techniques, in second with 28.8% said yes, In second place, 28.8% said yes, the landfill should receive any type of waste produced, 24.2% believe that, for the disposal of this waste should take into account the type of waste to be discarded, another 12.1% responded that batteries should be returned to the manufacturers for better management. Thus, the indicator evaluated is rated **FAVOURABLE**, since most of the population is aware that some materials require a more appropriate disposal site.

4.27. Adequate place for solid waste disposal according to the population

In Brazil, the form of management and disposal of solid waste by most Brazilian municipalities is considered inadequate, because its final disposal is conditioned á open air dumps, contributing to increased environmental impacts thus endangering the quality of life of the population. The same also points out that the lack of appropriate places for disposal is a factor that makes it even more difficult to solve this problem (COSTA et al., 2016).

It is important that the population claim improvements regarding the management of this waste, because according to Sousa, Ferreira, and Guimarães (2019) society has an important role in decision making regarding environmental issues, given that in current times the democratization of public policies aimed at the environment are more situated.

According to the interviewees' opinion, when asked if they would like the municipality to have an appropriate place for waste disposal, the participants were mostly in favor of the questioning, where 68.2% said yes, that the municipality should have an appropriate place for the packaging of these wastes, while 31.8% said no. This indicator is classified as **FAVOURABLE**, because the population longs for a suitable place for the treatment of solid waste.

4.28. Existence of dwellings near open air dumpsites

Bento (2020) points out that the garbage deposited near the residences of the population contributes to possible foci of disease transmitting agents besides allowing the contamination of the environment. Sousa, Ferreira, and Guimarães (2019) states that 50.8% of the waste produced in Brazil has as final destination, vacant lots, dumps, valley bottoms, etc.

Regarding the existence of residences near open air dumps, it was evidenced that 27.3% of the participants declared to live near this type of place, also pointing out that other nearby spaces would be similar to the aforementioned space. 72.7% answered that they did not live near these places. Considering that there is the presence of residences near these open-air places or similar, this indicator becomes **DISFavorable**, because there is a great possibility of the population to be contaminated with this waste nearby.

4.29. Knowledge by the population about reports of contamination due to contact with the waste.

According to Pinheiro, Costa et al. (2015) the leachate is considered the main factor for soil contamination, surface water and groundwater, having a high power of environmental impact, the same has been analyzed and studied for over 30 years. Costa et al. (2015) explains that leachate is a substance resulting from the mixture of organic and inorganic waste, becoming a complex substance with a high concentration of recalcitrant organic matter also possessing high toxic content. In addition, Bento (2020) states that there are open-air public spaces, other factors may contribute to possible sources of disease, such as the existence of animals, insects, the emergence of bad odor.

When asked about knowledge of cases of contamination due to contact with waste, 69.7% of respondents said they knew of cases that had occurred with people living near these sites, questioning that the diseases were from some type of bacteria acquired at these sites, while another 30.3% were unaware of any type of related case. Thus, this indicator is classified as **DEFAULT**.

4.30. Reports from residents about the presence of open-air roads near their residences.

According to Correia et al (2018), municipal dumps are spaces that represent a problem that affects all spheres, both economic, social, spatial, environmental and geopolitical. Thus, these spaces where waste is conditioned are considered illegal, where the same highlights that according to law 9.605/98 the implementation of open-air sites is an environmental crime, due to the damage it causes to the fauna and flora, in addition to pollution and contamination of soil and water.

According to the interviewees, 74.2% have heard neighbors, colleagues and even family members emphasize their dissatisfaction related to these disposal sites, because the reason would be the existence of bad odor and possible disease transmission agents affecting the quality of life of the people, another 25.8% said no. Thus the indicator evaluated is considered **DEFAULT**.

4.31. Knowledge of degraded areas that have been recovered in the municipality

For Ferreira et al (2019) degraded area is that place that somehow suffered with impacts triggered by some kind of disturbance being natural or being caused by activities carried out by humans, thus there is environmental damage and also the loss of regeneration of the fauna and flora of that locality

According to bento (2020) for there to be a recovery of degraded areas, there must be mainly removal and replacement of the entire damaged area, while using specialized techniques to facilitate the adaptation of fauna and flora in that new environment.

When questioned about their knowledge of degraded areas that have been recovered, most of the interviewees, 66.7%, stated that they did not know of any areas that had been recovered in any way by the public administration or the private sector. 33.3% declared that yes, some localities have been recovered, but they answered that their recovery was done with the purpose of expanding the urban area of the municipality. In this way it is classified as **DEFAULT**, since no public policy was implemented for the recovery of degraded areas.

4.32. Importance of distancing residences from this type of location

According to Klein, Gonçalves-Dias, and Jaya (2018), with the improper disposal of this waste, the environmental impacts caused can endanger public health and the quality of life of residents who may live near these sites, also causing environmental wear and soil and water contamination.

When asked about the importance of distancing their homes from open-air places as a way to mitigate the effects caused to the quality of life of the residents, it was found that 100% of the participants agree with the distancing of these places from their homes, thus it is classified as **FAVORABLE**, because the population knows the importance of the distance that allows better preservation of their health.

4.33. Waste Disposal in the Environment

When this waste is disposed of inappropriately in the environment, it suffers from the effects resulting from this disposal, with the contamination of the natural environment and loss of quality of life of the population (FERREIRA et al., 2019)

Oliveira, Santos, and Viana (2016) state that open-air public places harm the dynamics of the environment due to the possible contamination of the soil, water, and air, in addition to the presence of animals such as rats and vultures, which are vectors of agents that can transmit diseases.

In this question when asked about which resources could be harmed due to the disposal of this waste in an inappropriate   way in the environment, 60.61% responded that the health of the population would be the most affected, 56.06% stressed that it could compromise the fauna and flora of that location affected, 37.88% said that water would be affected, 15.15% that there would be the loss of air quality, another 24.24% stated the soil. This indicator is classified as **FAVOURABLE**, because the population recognizes that this waste affects the quality of all resources available to the population.

5. CONCLUSIONS

The study addressed is a problem that has long been discussed in the academic environment, where its relevance can be seen through questions associated with environmental degradation and quality of life of society. Given the relevance of the subject, which encompasses all aspects, especially public health and environmental degradation, it is necessary to discuss the final disposal of solid waste in the municipality of Uiraúna-PB and how they can affect the environment if there is no proper disposal. This work seeks to analyze and highlight points that can be improved with regard to the management of solid waste in the municipality aiming at a management that can benefit all involved.

The work in question included the application of a questionnaire with 37 questions for the population of the municipality of Uiraúna-PB, 4 questions were directed to trace a socioeconomic profile of the participants, and the other 33 focused on sustainability indicators. Regarding the questions related to sustainability indicators, twenty were considered as FAVORABLE and thirteen as DISFavorable, but it is important to note that most of the indicators that were highlighted as favorable are related to questions of knowledge and opinion about an adequate waste management and how these SUW could affect the population when not properly stored. While the unfavorable ones were focused on more relevant issues such as the inexistence of an appropriate place for the final disposal of waste, open air dumps near residences, and the contamination of the population through this waste.

After analyzing the data, they were transformed into quantitative data, and within the proposed margin, added up to 20 points, allowing us to consider the trend in the questionnaire as positive. Thus it could be seen that in fact the population knows that the municipality needs improvements in its waste management, and a place for its proper disposal, and the population is also in favor of public policies aimed at this issue.

Taking into account the degree of importance of the problem addressed, considering the environmental damage that this can lead, considering also all aspects of the natural environment and this includes fauna, flora, soil, air and water in addition to the loss of quality of life of the population, it was possible to reach some conclusions about the problem. The municipality in question has several problems regarding its management of urban waste which could be observed throughout the study, which highlighted the absence of an appropriate place for the disposal of urban waste, open air dumpsites near the homes of people, enabling a high risk of contamination to residents who live near these sites.

In this current context, the situation of the municipality goes against the law of National Policy on Solid Waste, so the indicator garbage collection and final disposal in open air dumps is considered DISFavorable to sustainability. Within this perspective it is clear the urgency that the municipality of Uiraúna-PB has, so that there are changes in their management of solid urban waste, so we can highlight strategies and methods that can minimize the impacts caused by mismanagement, as well as implement public policies aimed at environmental education, introduction of projects aimed at re-education with respect to the consumption of such waste and guidance on selective collection, and the installation of a landfill following all the rules and laws provided in the legislation, there may be the recovery of contaminated areas and their revitalization. Thus, only with the commitment of all involved, responsible agencies, social actors can we set goals and develop techniques and actions in order to solve the problem or mitigate its effects

REFERENCES

ABRELPE. **Solid Waste:** Manual of good practices in planning. Available at: <http://www.abrelpe.org.br/manual_apresentacao.cfm.> Accessed on: 03 April 2019.

BARBOSA, E. A. Resíduos sólidos: Aspectos conceituais e classificação. *In:* BARBOSA, E. A.; BATISTA, R. C.; BARBOSA, M. F. N. **Gestão dos recursos naturais:** uma visão multidisciplinar. Rio de Janeiro: Editora Ciência Moderna Ltda, 2012. p. 169-212.

BARBOSA, G. S. O desafio do desenvolvimento sustentável. **Revista Visões**, Rio de Janeiro, v. 1, n. 4, p. 1-11, jan./jun. 2008.

BARROS, J. D. de. S.; CHAVES, L. H. G.; FARLAS, S. A. R. Aspectos socioeconômicos da microbacia hidrográfica do Riacho Val Paraíso-PB. **Rev. Des. Regional**, Santa Cruz do Sul, v. 19, n. 1, p. 169-187, jan/apr. 2014.

BELLEN, H. M. V. **Indicadores de Sustentabilidade**. 1. ed. Rio de Janeiro: Editora FGV, 2005.

BENTO, F. F. L. **Evaluation of the management of urban solid waste through sustainability indicators in the municipality of Nazarezinho-PB**. 2020. 58 f. Monograph (Degree in Biological Sciences)-Federal University of Campina Grande, Cajazeiras-PB, 2020.

BERNARDO, M.; LIMA, R. da. S. Planning and implementation of a selective collection program: use of a geographic information system in the preparation of the routes. **Urbe, Revista Brasileira de Gestão Urbana,** Curitiba, v. 9, oct. 2017.

BERTICELLI, R.; PANDOLFO, A.; KORF, E. P. The integrated management of urban solid waste: perspectives and challenges. **Revista Gestão & Sustentabilidade Ambiental,** Florianópolis, v. 5, n.2, p. 711-744, oct. 2016./mar. 2017.

BRASIL. Complementary Law Nº 234/90 Urban Cleaning Code. **Chapter I of the preliminary provisions.** oct 1990. Available at: < http://lproweb.procempa.com.br/pmpa/prefpoa/dmlu/usu_doc/lei_comple mentar_234-90.pdf>. Accessed April 09, 2021.

BRAZIL. law no. 12.305, of august 2, 2010. National solid waste policy. 2. ed. Brasília: chamber of deputies, chamber editions, 2012.

CABRAL, J. V. B. et al, Hospital waste: the role of nursing in the health-disease process. **Revista Eletrônica Diálogos Acadêmicos**, v. 8, n. 1, p. 60-71, jan/jul. 2015.

CAFURE, V. A.; PATRIARCHA-GRACIOLLI, S. R. Health service waste and its environmental impacts: a literature review. **Interações**, Campo Grande, v. 16, n. 2, p. 301-314, jul/dec. 2015.

CAMARGO, M. E. et al. Solid waste from health services: a study on management. **Scientia Plena**, v. 5, n. 7, p. 1-14, mai/jul. 2009.

CARVALHO, V. G.; ESTENDER, A. C. Environmental awareness contributing to eliminate waste and expand the Actions in favor of the environment. **Revista Desafios**, Guarulhos-SP, v. 4, n. 2, p. 1-17, 2017.

CASTRO, A. L. C. **Application of Urban Solid Waste Sustainability Indicators in the municipality of Uberlândia-MG**. 2016. 69 f. Course Conclusion Paper (Bachelor's Degree in Environmental Engineering) - Universidade Federal de Uberlândia, Uberlândia, 2016.

CAVALCANTI, M. C. C. **Environmental awareness campaign: discursive and social practice in late modernity**. 2013. 266 f. Thesis (Doctorate in Literature)- Universidade Federal de Pernambuco, Recife-PE, 2013.

UNCED - UNITED NATIONS CONFERENCE ON ENVIRONMENT AND DEVELOPMENT. **Agenda 21.** 1996. Available at: <www.mma.gov.br/responsabilidade-socioambiental-agenda-21/agenda-21-global>. Accessed 04 April. 2019.

CEMPRE - BUSINESS COMMITMENT FOR RECYCLING. **Municipal Waste: Manual for Integrated Management.** 2010.

CIQUETTI, H. S. Lixo, resíduos sólidos e reciclagem: uma analise comparativa dos recursos didáticos. **Editora UFPR**, n. 23, p. 307-333, 2004.

CONDE, T. T.; STACHIW, R.; FERREIRA, E. Landfill as an alternative for environmental preservation. **Brazilian Journal of Amazonian Sciences**, v. 3, n. 1, p. 69-80, 2014.

CORREIA, V. M. S. et al. Case study: aspects and perceived impacts on the location of municipal dumpsites using the M-MACBETH tool. **DAE Journal**, N. 211, v. 66, Jul/Sept. 2018.

COSTA, F. M. da. et al. Treatment of leachates from solid waste landfills using fenton and solar photo-Fenton processes. **Rev. Ambient. Water**, v. 10, n. 1. p. 107-116, 2015.

COSTA, T. G. A. et al. Environmental impacts of open air dumpsite in the municipality of Cristalândia, Piauí State, northeastern Brazil. **Rev. Bras. Gest. Sustent**, V. 3, n. 4, p. 79-86, mai/jun. 2016.

CPRM - Geological Service of Brazil. **Projeto cadastro de fontes de abastecimento por água subterrânea Estado da Paraíba:** diagnóstico do Município de Uiraúna. Recife: CPRM/PRODEEM, 2005.

CUNHA, V.; FILHO, J. V. C. Gerenciamento da coleta de resíduos sólidos urbanos: estruturação e aplicação de modelo não-linear de programação por metas. **Gestão e Produção**, v. 9, n. 2, p. 143-161, august 2002.

FERREIRA, R. S. Et al. Degraded areas: environmental reforestation techniques. **Revista cientifica multidisciplinar núcleo de conhecimento**, v.11, p. 71-84, jun. 2019.

GIACCOM-RIBEIRO, M. B.; MENDES, C. A. B. Evaluation of parameters in the estimation of urban solid waste generation. **Brazilian journal of development planning**. v. 7, n. 3, p. 422-443, Aug. 2018.

GOMES, L. P. et al. Avaliação ambiental de aterros sanitários de resíduos solidos urbanos procedidos ou não por unidades de compostagem. **Engent Sanit Ambient**, São Leopoldo, v. 20, n. 3, p. 449-462. jul/set. 2015.

GONÇALVES, S. A. The National Policy for Solid Waste: some notes on Law 12.305/2010. In: *SANTOS, M. C. L. S.; DIAS, S. L. F. G.* **Resíduos sólidos Urbanos e seus impactos socioambientais**. São Paulo: IEE-USP, 2012. p. 40-47.

GOUVEIA, N. Resíduos sólidos urbanos: impactos socioambientais e perspectiva de manejo sustentável com inclusão social. **Ciência & saúde coletiva, v.**17, n.6, p.1503-1510, 2012.

IBGE- BRAZILIAN INSTITUTE OF GEOGRAPHY AND STATISTICS. *Available at:* < https://cidades.ibge.gov.br/brasil/pb/uirauna/panorama>. Accessed on: 27 March 2020.

KEMERICK, P. D. da. C.; RITTER, L. G.; BORBA, W. F. Environmental sustainability indicators: methods of applications. **Revista de Monografias Ambientais - REMOA**, Santa Maria: v. 13, n. 5, p. 3723-3736, 2014.

KLEIN, F. B,; GONÇALVES-DIAS, S. L. F,; JAYA, M. Urban solid waste management in the upper Tietê watershed: an analysis on the use of ICT in accessing government information. **Brazilian journal of urban management**. *São Paulo, p. 140-153, jan/apr. 2018.*

LEITE, M. F. **A taxa de coleta de resíduos sólidos domiciliares uma análise critica**. 2006. 106 f. Dissertation (Master in Civil Engineering) - Engineering School of São Carlos, Federal University of São Paulo. São Carlos, 2006.

LIMA, C. C. T. **Application of Urban Solid Waste Sustainability Indicators in the Municipality of Araguari-MG.** 2017. 61f. Course

Conclusion Paper (Bachelor in Environmental Engineering) - Institute of Agrarian Sciences, Federal University of Uberlândia - UFU, Uberlândia. 2017.

LIMA, T. da. S. **Evaluation of solid waste management through pressure-state-impact-response (PEIR) sustainability indicators in the municipality of Cachoeira dos Indios - PB**. 2018. 44 f. Monograph (Degree in Biological Sciences)-Federal University of Campina Grande, Cajazeiras-PB, 2018.

MANCINI, S. D.; FERRAZ, L. J.; BIZZO, W. A. Resíduos sólidos. In: ROSA, A. H.; FRACETO, L. F.; MOSCHINI-CARLOS, V. **Meio Ambiente e sustentabilidade**. Porto Alegre: Bookman, 2012. p. 346-374.

MARCO, E. de. **Study of the integrated management plan of solid waste in the municipality of Cotiporã-RS**. 2014. 90 f. Monograph (Bachelor in Environmental and Sanitary Engineering)-Federal University of Pelotas, Pelotas, 2014.

MATOS, B. B. de. M. **Study of the reuse, recycling and final destination of civil construction waste in the city of Rio de Janeiro**. 2013. 83 f. Monografia (Bacharel em Engenharia)-Universidade Federal do Rio de Janeiro, Rio de Janeiro, 2013.

MEDEIROS, G. A. de.; GIORDANO, L. do. C.; REIS, F. A. G. V. Environmental management. In: ROSA, A. H.; FRACETO, L. F.; MOSCHINI-CARLOS, V. **Meio ambiente e sustentabilidade**. Porto: Bookman, 2012. p. 375-407.

MUCELIN, C.; A. BELLINI. M. Trash and perceived environmental impacts on the urban ecosystem. **Sociedade & Natureza, v.** 20, n. 1, p. 111-124, 2008.

NETO, P. N.; MOREIRA, T. A. Urban solid waste management in the Metropolitan Region of Curitiba: regional composting policy. **Revista Geografar**, Curitiba, v.4, n.2, p.72-96, jul./dez. 2009.

OLIVEIRA, K. C,; SANTOS, R. M. da. S.; VIANA, A. L. Solid waste generation: the perception of the population in a city neighborhood of Manaus, Amazonas. **InterfaceHS**, São Paulo, v. 11, n. 1, p. 1-11 jun. 2016.

OLIVEIRA, M. C. B. R. de. **Management of post-consumer plastic waste for recycling in Brazil**. 2012. 104 f. Dissertation (Master in Energy Planning)- Universidade Federal do Rio de Janeiro, Rio de Janeiro, 2012.

PEREIRA, S. S.; CURI, R. C.; CURI, W. F. Use of indicators in the management of municipal solid waste: a methodological proposal for construction and analysis for municipalities and regions. **Eng Sanit Ambient**, Campina Grande-PB, v. 23, n. 3, p. 471-483, may./jun. 2018.

PORDEUS, A. V.; BARROS, J. D. de. S. et al. Socioeconomic aspects of the várzeas de Sousa irrigated perimeter (PIVAS) in the semi-arid Paraibano. **Pesquias de ensino em ciências exatas e da natureza**, Cajazeiras-PB, 2019.

PORTELA, M. O.; RIBEIRO, J. C. J. Aterros sanitários: aspectos gerais e destino final dos resíduos. **Revista direito ambiental e sociedade**. Belo horizonte. v. 4, n. 1, p. 115-134, 2014.

PRODANOV, C. C.; FREITAS, E. C. de. **Metodologia do trabalho cientifica**: métodos e técnicas do trabalho acadêmico. Novo Hamburgo: Feevale, 2013.

QUEIROZ, A. J. P. **Perception of the population about the urban solid waste in the context of basic sanitation in the municipality of Barra de São Miguel-PB**. 2011. 60 f. Monograph (Bachelor in Sanitary and Environmental Engineering)- Universidade Estadual da Paraiba, Campina Grande-PB, 2011.

QUERINO, L. A.L.; PEREIRA, J. P. I.G. Geração de resíduos sólidos: percepção da população de São Sebastião de lagoa de roça, Paraíba. **Revista monografias ambientais**. v. 15, n. 1, p. 404-415, jan/abr, 2016.

RICHTER, L. T. **Importance of awareness and selective collection in the municipality of Palmito - SC**. 2014. 78 f. Monograph (Specialist in Environmental Management) - Federal Technological University of Parana, Medianeira, 2014.

RODRIGUES, C. R. P.; MENTI, M. de. M. Solid waste: management and federal public policies. **Caderno do programa de direito de pós-graduação de direito PPGDir/UFRGS**. v. 11, n. 3, p. 59-79, 2016.

RODRIGUES, R. R. **Reduction of environmental impacts caused by urban solid waste through a selective collection**. 2010. 33 f. Course completion paper (Graduation in Biological Sciences) - Course of Biological Sciences, Methodist University Center Isabela Hendrix, Belo Horizonte, 2010.

SANTOS, M. C. L. do.; DIAS, S. L. F. G. **Resíduos sólidos urbanos e seus impactos socioambientais**. São Paulo: IEE-USP, 2012.

SANTOS, M. C. L.; GONÇALVES-DIAS, S. L. F.. Waste management in the city of São Paulo: one problem, multiple solutions. 2012. In: PADOVANO, B.R.; NAMUR, M.; SALA, P. B. (Eds.), **São Paulo:** em busca da sustentabilidade. São Paulo: EDUSP/PINI. 2012. p. 146-159.

SHMITZ, M. **Management of domestic solid waste: a case study in the sorting center, treatment and final destination of domestic solid waste in the municipality of Estrela-RS.** 2012. 78 f. Monograph (Bachelor in Environmental Engineering)- Univates University Center, Lajeado, 2012.

SILVA, E. da. **Environmental education: urban waste from problem to possibility.** 2015. 22 f. Monografia (Specialization in Education in Human Rights)- Universidade Federal do Paraná, Paranaguá-PR, 2015.

SILVA, E. L. da.; MENEZES, E. M. A pesquisa e suas classificações. In: SILVA, E. L. da.; MENEZES, E. M. **Metodologia da pesquisa e elaboração de dissertação.** 4. ed. Florianópolis: UFSC, 2005. p. 19-23.

SILVA, N. M. da.; NOLÊTO, T. M. S. J. Reflexões sobre lixo, cidadania e consciência ecologia. **Revista eletrônica do curso de geografia do campus avançado de Jataí-GO**, Jataí-GO, n. 2, p. 1-14, jan/jun. 2004.

SINGH, R. K.; MURTY, H. R.; GUPTA, S. K.; DIKSHIT, A. K. Development of composite sustainability performance index for steel industry. **Ecological Indicators**, v. 7, n. 3, p. 565-588, 2007.

SIQUEIRA, M. M.; MORAES, M. S. de. Saúde coletiva, resíduos sólidos urbanos e os catadores de lixo. **Cien. Saúde coletiva**. Rio de Janeiro. v. 14, n. 16, dec. 2009.

SOARES, A. M. **Evaluation of solid waste management through the system of sustainability indicators pressure-state-impact-response (PEIR) in the municipality of Nazarezinho-PB**. 2017. 87 f. Monograph (Degree in Biological Sciences)- Universidade Federal de Campina Grande, Cajazeiras-PB, 2017.

SOUSA, G. L. M,; FERREIRA, V. T. de. O,; GUIMARÃES, J. de. C. Open air dumpsite: implications for the environment and society. **Revista valores**. Volta Redonda. p. 367-376, 2019.

TENORIO, J. A. S.; ESPINOSA, D. C. R. Controle Ambiental de Resíduos. In: PHILIPPI JR, A, ROMÉRO MA, BRUNA GC (Org.). **Environmental Management Course**. Barueri SP: Manoela, 2004. p. 155-213.

VASQUEZ, S. F.; BARROS, J. D. de S. ; SILVA, M. de F. P da. Alternatives to conventional agriculture. **Revista Verde de Agroecologia e Desenvolvimento Sustentável**, Mossoró, v.3, n. 3, p. 06-12, jul./set. 2008.

VICENTE, A. R. P.; BERTOLINI, G. R. F.; RIBEIRO, I. Reception of the population regarding the sustainability indicators of Curitiba: the sustainable city of the planet. **R. Gest. Sust. Ambient**, Florianópolis, v. 4, n. 2, p. 254-272, oct/mar. 2016.

VIEIRA, A. C. P.; GARCIA, J. R. A gestão de resíduos sólidos domésticos no Brasil a par da experiência internacional. **Revista Economia e Tecnologia (RET)**, v. 8, n. 4, p. 57-66, oct/dec. 2012.

ZANELLA, L. C. H. **Metodologia de pesquisa**. Florianópolis: University of Administration Sciences/UFSC, 2013.

ZASSO, M. A. de. C. et al. **Environment and sustainability**. Ijuí: Editora Unijuí, 2014.

APPENDIX

<table>
<tr>
<td></td>
<td>

FEDERAL UNIVERSITY OF CAMPINA GRANDE

TEACHER TRAINING CENTER

ACADEMIC UNIT OF EXACT AND NATURAL SCIENCES

COURSE: LICENTIATE IN BIOLOGICAL SCIENCES

</td>
</tr>
</table>

Questionnaire

1. Gender: Male (   )   Female (   )

2. Age: (   ) ≤ 20   (   ) 20 – 30   (   ) 30 – 40   (   ) >= 40

3. Level of education:

   (   ) Illiterate

   (   ) Incomplete elementary school

   (   ) Elementary School Complete

   (   ) Incomplete High School

   (   ) High School Complete

   (   ) Higher Education Incomplete

   (   ) Higher Education Complete

4. Number of people in the family:

   1 (   )   2 (   )   3 ( )   (   ) 5 or more

5. Do you have any knowledge about what is Urban Solid Waste?

( ) Yes   ( ) No

6. The largest amount of garbage in your city is?

( ) organic   ( ) inorganic

7. What types of garbage are produced in your municipality:

( ) Domiciliary   ( ) Pruning   ( ) Commercial   ( )
Agricultural

( ) Debris   ( ) Public   ( ) Industrial

8. Who performs the urban cleaning service?

( ) City Hall   ( ) Others   ( ) City Hall and others

9. Does collection reach the entire urban area of the municipality?

( ) yes   ( ) no

10. How often is the domestic collection being done?

( ) Daily   ( ) Alternate   ( ) Weekly   (   )
Occasional

11. Does the city charge for the urban cleaning service?

( ) Yes

( ) Rate

( ) Tariff

( ) Together with IPTU

(   ) Other. Specify

____

(   ) No

12. How satisfied are you with the garbage collection in your city?

(   ) Great   (   ) Good   (   ) Fair   (   ) Poor

13. What kind of waste do you and your family produce at home?

(   ) paper

(   ) Plastic

(   ) Organic material (fruits, vegetables...)

(   ) Other, which ____________________

14. Do you know what selective collection is?

(   ) Yes   (   ) No

15. How do you and your family usually store this waste at home?

( ) in a plastic garbage can.

( ) In plastic bags.   ( ) Other. How?

____

16. Do you separate your Household Waste?

(   ) Yes   (   ) No

17. Do you reuse the waste from your residence?

(   ) Yes   (   ) No

18. Would you separate your household waste for recycling in your municipality?

( ) Yes   ( ) No

19. You know where the waste you produce goes.

( ) to the dump

( ) to landfill

( ) to the controlled landfill

( ) for incineration

( ) Does not know what it is. ( ) Other. Which one?

____

20. When you go shopping, do you worry about choosing products that are less harmful to the environment?

( ) Yes ( ) No

21. Is there selective collection in your municipality?

( ) Yes   ( ) No

22. How satisfied are you with the selective collection in your city?

( ) Great   ( ) Good   ( ) Fair   ( ) Poor

23. Do you know if hospital waste such as syringes, gloves, and batteries receive a different destination than the rest of the urban waste?

(   ) Yes, which? _______________   (   ) No

24. Are you aware of the health hazards that this waste (hospital waste and batteries) represents to the environment and to the population's health?

(   ) Yes   (   ) No

25. Would you like the public agencies responsible for waste management to conduct environmental awareness campaigns aimed at the population?

(   ) Yes   (   ) No

26. How would you like to receive information about waste and its impact on the environment?

(   ) Orientation visit

(   ) Pamphlets or Posters

(   ) Community meetings

(   ) Radio Information

(   ) Information in newspaper

(   ) Other, Which? _______________

(   ) Wouldn't like to

27. For you, what are the main problems caused by the incorrect disposal of garbage in your city?

(population contracting diseases   ( ) Loss of water quality

environment and its resources

( ) Generation of bad odor   ( ) none of the alternatives

28. Who do you believe is mainly responsible for the damage caused to the environment in your municipality?

( ) Agricultural   sector ( ) Commercial sector

( )Municipal Government   ( ) Population

( ) Other, which? __________   ( ) Industrial sector

29. Do you consider the current final destination of the Urban Solid Waste in your city to be the most appropriate?

( ) Yes   ( ) No

30. What would your opinion be if a sanitary landfill system was implemented in your city?

( ) Excelent   ( ) Good   ( ) Regular   ( ) Bad

31. In your opinion, should the landfill receive all kinds of garbage?

( ) Yes, because it was made for this.

( ) Yes, as long as you know which garbage can be thrown in the landfill.

( ) No, because hospital waste needs to go somewhere else.

( ) No, because materials such as batteries need to go back to the companies and have a different destination.

32. For you, garbage disposal in open areas compromises the quality of life:

( ) soil   ( ) air   ( ) water   ( ) plants and animals of that location

( ) Population nearby   ( ) none of the alternatives

33. Do you live near open-air roads?

( ) Yes   ( ) No

34. Have you, someone in your family, or an acquaintance ever become ill due to living near improper waste disposal sites in your city?

( ) Yes   ( ) No

35. Have you heard reports of dissatisfaction from anyone living near these places?

( ) Yes   ( ) No

36. Do you know of any areas degraded by Solid Urban Waste management that have already been recovered?

( ) Yes   ( ) No

37. Do you think it is important to build these final waste disposal sites away from residential areas, aiming at the safety and health of the population?

( ) Yes   ( ) No

ANNEXES

ANNEX A. Declaration of project approval by CEP

UNIVERSIDADE FEDERAL DE CAMPINA GRANDE
CENTRO DE FORMAÇÃO DE PROFESSORES
COMITÊ DE ÉTICA EM PESQUISA

DECLARAÇÃO

Declaro para os devidos fins que, o projeto de pesquisa intitulado: **"ANÁLISE DO GERENCIAMENTO DE RESÍDUOS SóLIDOS URBANOS ATRAVÉS DE INDICADORES DE SUSTENTABILIDADE NO MUNICÍPIO DE UIRAÚNA-PB"**, com o Certificado de Apresentação para Apreciação Ética - CAEE, nº: 15390019.0.0000.5575, sob responsabilidade do professor José Deomar de Souza Barros, foi aprovado pelo Comitê de Ética em Pesquisa - CEP do Centro de Formação de Professores da Universidade Federal de Campina Grande, em agosto de 2019.

Cajazeiras-PB, 08 de Abril de 2021

Prof. Dr. Paulo Roberto de Medeiros
Coordenador do CEP/CFP/UFCG
Mat. SIAPE Nº 1965184

Rua Sérgio Moreira de Figueiredo, s/n – Casas Populares – CEP 58900-000 Cajazeiras / PB
Telefone: (83) 3532-2074 / Endereço eletrônico: cepcfpufcgcz@gmail.com

More
Books!

OMNIScriptum

Printed by Books on Demand GmbH, Norderstedt / Germany